I0814058

SCIENCE QUESTIONS

HOW DO FISH BREATHE?

Cody Koala
An Imprint of Pop!
popbooksonline.com

by Elizabeth Andrews

abdobooks.com
Published by Pop!, a division of ABDO, PO Box 398166, Minneapolis, Minnesota 55439.

Printed in the United States of America, North Mankato, Minnesota

102021
012022

THIS BOOK CONTAINS RECYCLED MATERIALS

Cover Photo: Shutterstock Images
Interior Photos: Getty Images, 6; Shutterstock Images 5, 9 10, 13, 15, 16, 18, 21

Editor: Grace Hansen
Series Designer: Laura Graphenteen

Library of Congress Control Number: 2021942247

Publisher's Cataloging-in-Publication Data
Names: Andrews, Elizabeth, author.
Title: How do fish breathe? / by Elizabeth Andrews
Description: Minneapolis, Minnesota : Pop!, 2022 | Series: Science questions | Includes online resources and index
Identifiers: ISBN 9781098241070 (lib. bdg.) | ISBN 9781098241773 (ebook)
Subjects: LCSH: Fishes--Juvenile literature. | Fishes--Respiration--Juvenile literature. | Gills--Juvenile literature. | Fishes--Behavior--Juvenile literature. | Children's questions and answers--Juvenile literature.
Classification: DDC 500--dc23

Hello! My name is

Cody Koala

Pop open this book and you'll find QR codes like this one, loaded with information, so you can learn even more!

Scan this code* and others like it while you read, or visit the website below to make this book pop.

popbooksonline.com/fish-breathe

*Scanning QR codes requires a web-enabled smart device with a QR code reader app and a camera.

Table of Contents

Chapter 1

Something in the Water

Fish need **oxygen** to live, just like humans and land animals. But fish live underwater. They can't get oxygen from the air. Instead, fish get oxygen from water.

Watch a video here!

Water and air are made up of tiny things called **molecules**. There are different kinds of molecules in water.

Oxygen is one of them. It **dissolves** into water from the air. Wind, rain, **currents**, and waves mix in the oxygen.

Chapter 2

The Gills

Fish take in water through their mouths. Water then flows back to the gills. Gills are slits on the side of a fish's head. They are filled with many tiny red blood **vessels** called capillaries.

gills

Learn more here!

gills

Gills need to be protected. Strong but flexible bony plates called operculum lay over the gills. The capillaries are also supported by bony loops called gill arches.

Fish gills open and close to push water out after the **oxygen** has been taken in.

As the blood flows through the capillaries, it takes in the oxygen **molecules** from the water. The oxygen attaches to the blood. Then it is carried through the fish's body to its **cells**.

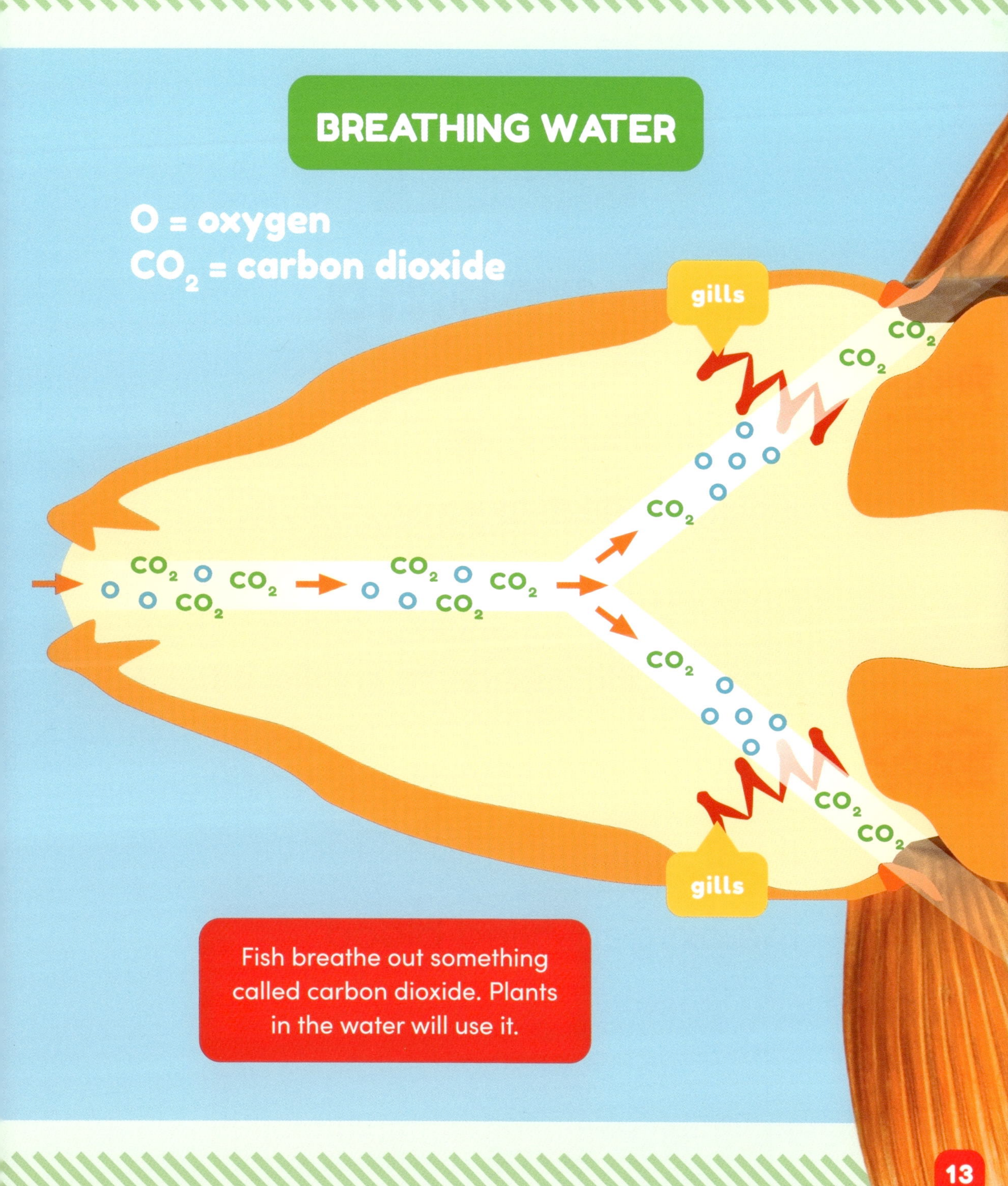
BREATHING WATER
O = oxygen
CO_2 = carbon dioxide
gills
gills
Fish breathe out something called carbon dioxide. Plants in the water will use it.

Chapter 3

Pumping Oxygen

Bodies need **oxygen**. Oxygen gives **cells** what they need to work. So, it needs to make its way to every cell in the body. It is the heart's job to move the oxygen-filled blood.

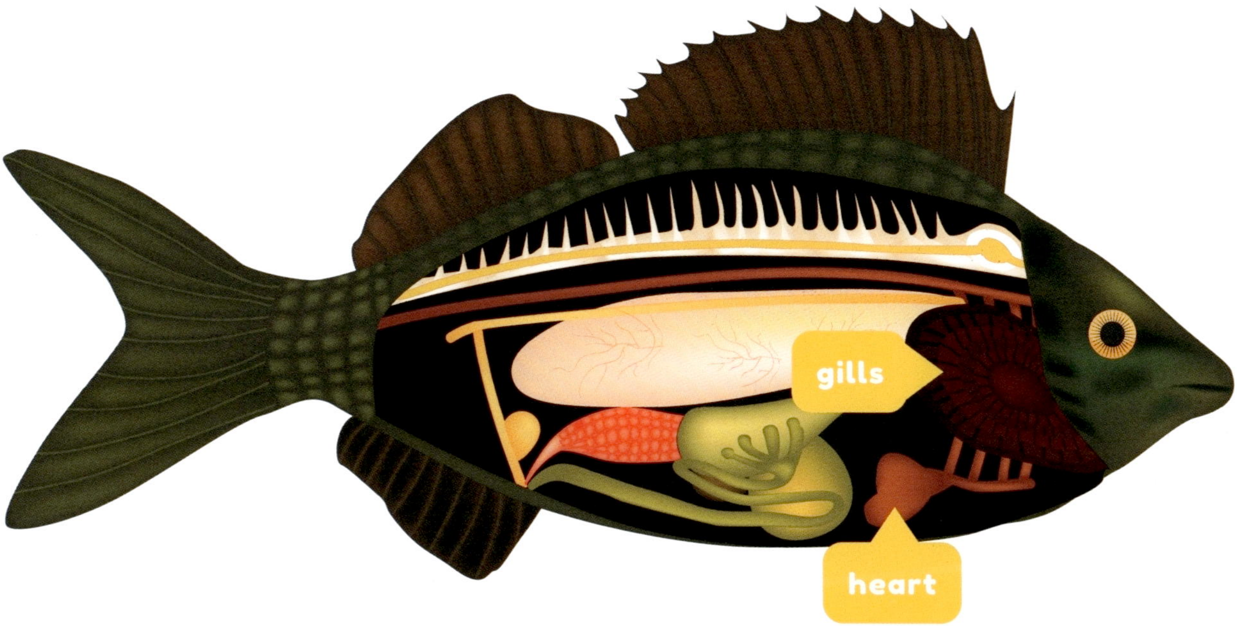

gills

heart

Water is thicker than air. It also has less oxygen in it than air.

This means that it takes more work for a fish to get the oxygen it needs.

capillaries

Since water is low in oxygen, fish had to adapt. The capillaries in the gills have a lot of little folds. These give fish more chances to take in the oxygen in the water.

Chapter 4

Changing Waters

Scientists watch **oxygen** levels in water. They noticed that levels in the ocean are decreasing. **Climate change** might be the cause of this. Scientists are working on ways to slow down the fall of oxygen levels.

Complete an activity here!

Making Connections

Text-to-Self

Have you ever seen fish gills up close? Where were you when you saw them?

Text-to-Text

Have you read any other books about fish? Did you learn anything that wasn't in this book?

Text-to-World

Fish breathe with their gills. What parts of the body help humans breathe?

Glossary

cell – the simplest unit of life that makes up all living things.

climate change – a long-term change in Earth's climate. It includes changing temperatures and weather patterns. It can result from natural processes or human activities.

current – a flow of water.

dissolve – to become liquid.

molecule – the smallest part of a material.

oxygen – a colorless, tasteless gas in the air. It is needed to live.

vessel – a tube that blood moves through.

Index

Online Resources

popbooksonline.com

Thanks for reading this Cody Koala book!

Scan this code* and others like it in this book, or visit the website below to make this book pop!

popbooksonline.com/fish-breathe

*Scanning QR codes requires a web-enabled smart device with a QR code reader app and a camera.